PARTS OF A PLANT

ROOTS

A Crabtree Roots Book

ALICIA RODRIGUEZ

Crabtree Publishing
crabtreebooks.com

School-to-Home Support for Caregivers and Teachers

This book helps children grow by letting them practice reading. Here are a few guiding questions to help the reader with building his or her comprehension skills. Possible answers appear here in red.

Before Reading:

- What do I think this book is about?
    - *This book is about roots.*
    - *This book is about what roots look like.*

- What do I want to learn about this topic?
    - *I want to learn what roots are.*
    - *I want to learn what roots look like.*

During Reading:

- I wonder why...
    - *I wonder why some roots grow in soil.*
    - *I wonder why some roots are long.*

- What have I learned so far?
    - *I have learned that some roots grow in water.*
    - *I have learned that some roots are big.*

After Reading:

- What details did I learn about this topic?
    - *I have learned that roots can be different sizes and lengths.*
    - *I have learned that most plants have roots.*

- Read the book again and look for the vocabulary words.
    - *I see the word **soil** on page 8 and the word **plants** on page 13. The other vocabulary word is found on page 14.*

These are **roots**.

Some roots are big.

Some roots are long.

Some roots are
in **soil**.

Some roots are in water.

Most **plants**
have roots!

Word List

Sight Words

are
big
have
in
most
some
these
water

Words to Know

plants

roots

soil

25 Words

These are **roots**.

Some roots are big.

Some roots are long.

Some roots are in **soil**.

Some roots are in water.

Most **plants** have roots!

PARTS OF A PLANT
ROOTS

Written by: Alicia Rodriguez
Designed by: Rhea Wallace
Series Development: James Earley
Proofreader: Ellen Rodger
Educational Consultant: Marie Lemke M.Ed.

Photographs:
Shutterstock: showcake: cover; ifong: p. 1; Thammanoon Khamchalee: p. 3, 14; siambizkit: p. 5; Noppanunk: p. 6-7; ER_09: p. 9; Damsea: p. 11; Ian Grainger: p. 12, 14

Crabtree Publishing

crabtreebooks.com 800-387-7650

Printed in Canada/032024/CP20240312

Published in Canada
Crabtree Publishing
616 Welland Avenue
St. Catharines, Ontario
L2M 5V6

Published in the United States
Crabtree Publishing
347 Fifth Avenue
Suite 1402-145
New York, NY 10016

Hardcover	978-1-4271-4067-8
Paperback	978-1-4271-4073-9
Ebook (pdf)	978-1-4271-3331-1
Epub	978-1-4271-3391-5
Read-along	978-1-4271-4079-1
Audio book	978-1-4271-3913-9

Library and Archives Canada Cataloguing in Publication

Title: Roots / Alicia Rodriguez.
Names: Rodriguez, Alicia (Children's author), author.
Description: Series statement: Parts of a plant | "A Crabtree roots book".
Identifiers: Canadiana (print) 20210176059 | Canadiana (ebook) 20210176067 | ISBN 9781427140678 (hardcover) | ISBN 9781427140739 (softcover) | ISBN 9781427133311 (HTML) | ISBN 9781427133915 (EPUB) | ISBN 9781427140791 (read-along ebook)
Subjects: LCSH: Roots (Botany)—Juvenile literature.
Classification: LCC QK644 .R63 2022 | DDC j575.5/4—dc23

Library of Congress Cataloging-in-Publication Data

Names: Rodríguez, Alicia, 1992- author.
Title: Roots / Alicia Rodriguez.
Description: New York, NY : Crabtree Publishing Company, [2022] | Series: Parts of a plant- a Crabtree roots book | Includes index.
Identifiers: LCCN 2021014290 (print) | LCCN 2021014291 (ebook) | ISBN 9781427140678 (hardcover) | ISBN 9781427140739 (paperback) | ISBN 9781427133311 (ebook) | ISBN 9781427133915 (epub) | ISBN 9781427140791
Subjects: LCSH: Roots (Botany)--Juvenile literature. | Plants--Juvenile literature.
Classification: LCC QK644 .R63 2022 (print) | LCC QK644 (ebook) | DDC 581.4/98--dc23
LC record available at https://lccn.loc.gov/2021014290
LC ebook record available at https://lccn.loc.gov/2021014291